Murilo O. Fernandes
Otacilio J. P. Rangel
Jéferson L. Ferrari

Precision Agriculture

Murilo O. Fernandes
Otacilio J. P. Rangel
Jéferson L. Ferrari

Precision Agriculture

Basic principles of precision agriculture applied to coffee plantations

ScienciaScripts

Imprint
Any brand names and product names mentioned in this book are subject to trademark, brand or patent protection and are trademarks or registered trademarks of their respective holders. The use of brand names, product names, common names, trade names, product descriptions etc. even without a particular marking in this work is in no way to be construed to mean that such names may be regarded as unrestricted in respect of trademark and brand protection legislation and could thus be used by anyone.

Cover image: www.ingimage.com

This book is a translation from the original published under ISBN 978-613-9-63352-4.

Publisher:
Sciencia Scripts
is a trademark of
Dodo Books Indian Ocean Ltd. and OmniScriptum S.R.L publishing group

120 High Road, East Finchley, London, N2 9ED, United Kingdom
Str. Armeneasca 28/1, office 1, Chisinau MD-2012, Republic of Moldova, Europe
Printed at: see last page
ISBN: 978-620-7-65891-6

ACKNOWLEDGEMENTS

This book would not have been possible without the help, encouragement and commitment of several people. I would therefore like to express my gratitude and appreciation to all those who, directly or indirectly, helped make this task a reality. First and foremost, my friend and professor Dr Otacilio José Passos Rangel, for the efforts of his guidance, the usefulness of his recommendations and the warmth with which he always welcomed me. I would like to thank him immensely for the freedom of action he allowed me, which was decisive in bringing this book to its final version. To my boyfriend, Felipe Ferreira, for his unconditional patience, unrestricted availability and his demanding, critical and creative way of arguing, which made it easier for me to achieve my professional and personal goals. To my friends who have never been absent, I thank them for their friendship and affection. To my mother, for the solid education she gave me, which allowed me to continue in my academic life, my eternal thanks. To everyone, thank you.

CONTENTS

PREFACE

Coffea conefora, better known as conilon coffee, is the most widely grown coffee species in the world. In Brazil, more specifically in the state of Espírito Santo, this species is of significant importance in the state's socio-economic scenario and makes a strong contribution to the national economy, making Brazil the world's largest producer of the species. As well as boosting the national economy, the fruit has been considered a functional food that not only nourishes, but also maintains health by demonstrating anticarcinogenic, anti-inflammatory and antioxidant effects, contributing to the well-being of those who consume it. In order to have a good production that is sufficient to meet the producer's demand and continue to contribute to the economy without waste and, above all, without further affecting the environment, some care is needed in coffee farming. Precision farming is an innovative method that makes coffee cultivation more conscious and economical, without the indiscriminate use of correctives, fertilisers and pesticides that damage the soil and the habitat of various species. In view of the importance of consumption and especially the cultivation of conilon coffee, this book is subdivided into six chapters that explain everything from the history of coffee cultivation in Brazil, description such as taxonomy, classification and general characteristics of the plant and economic importance. In addition to a simplified demonstration of soil preparation (fertilisation and liming), nutrition (the role of each nutrient), a detailed demonstration of the technique of precision agriculture, with detailed methodology from the collection of material to the technical and statistical analyses through a complete case study followed by a discussion.

INTRODUCTION

Coffea canephora is the second most cultivated species of the genus worldwide, accounting for around 35 to 38 per cent of coffee production. Espírito Santo stands out as the largest Brazilian producer of this species, popularised by the name conilon coffee, which began to be grown in the state around 1912 (FERRÃO et al., 2007). The name 'Conilon' originated from the word Kouillou with the letters K and U replaced by C and N, respectively (FAZUOLI, 1986). The frost of 1975 and the eradication of coffee plantations were historic milestones for the species in Espírito Santo, sparking great interest in the species and decisively marking the history of production in the state.

Conilon coffee (Coffea canephora) occupies a space of great importance in the socio-economic scenario in the state of Espírito Santo, and its cultivation, in particular, is growing and significant for producers. The cultivation of coffee also continues to heat up the national economy, giving the country the profile of the world's largest producer (ENCARNAÇÃO and LIMA, 2003).

According to the Brazilian Coffee Industry Association - popularly known as ABIC - between November 2014 and December 2015 domestic consumption of the beverage totalled around 20.5 million bags, an increase of 0.86% on the previous survey. According to ABIC, 36,890,470 60kg bags of coffee were exported from Brazil in 2015, including its three main states of consumption: raw, roasted and soluble.

The flavours and aromas present in the fruit are the factors responsible for making it so inviting and for making the consumption of its beverage an increasingly constant and common habit for all of humanity (MOREIRA et al., 2001). According to recent studies (NACIF, 2003), coffee has also proved to be a potentially functional food.

As SANDERS (1998) rightly points out, "functional" foods are those which, as well as nourishing, have the potential to maintain health. It is worth emphasising, however, that the effects of their possible consumption have nothing to do with possible cures for a particular disease, but rather with the qualitative promotion of health and well-being for individuals.

Caffeine and the chlorogenic acids present in its composition have also been the basis for a variety of scientific publications on the pharmacological effects of regular coffee consumption. In their work on the subject, DEL CASTILLO, AMES and GORDON (2002) describe that melanoidins are formed during the roasting process by the Maillard reaction and part of the chlorogenic acids are incorporated, which, once consumed, can bring some benefits to human health.

Among these benefits are the occurrence of some biological processes, such as inhibition of leukotriene biosynthesis, as well as anticarcinogenic, anti-inflammatory and antioxidant effects (BALASUBACHINI et al., 2004). Through in vivo and epidemiological studies, CINTRA and MANCINI FILHO (2001) found an inverse relationship between the consumption of antioxidant foods and the incidence of chronic non-communicable diseases.

Due to their ability to delay and even inhibit lipid oxidation, the incidence of antioxidants in industrialised products and the occurrence of foods that contain natural substances with the same effect - such as coffee - have further influenced the consumer market's choices when it comes to the shelf (SILVA and NAVES, 2001).

Whether it's because of its flavour and aroma or because of its antioxidant potential, there are more than enough elements in coffee to justify the demand of a demanding consumer market and, likewise, the relevance of scientific research dedicated to analysing this fruit. In the same vein, there is also the importance of developing academic work that is concerned with a particular aspect of its production, the handling of which even precedes storage, the chemical process that makes it soluble and even the roasting of the beans. In short, it's the soil.

Soil is understood to be any body of unconsolidated material that covers the earth's surface lying between the lithosphere and the atmosphere (SCHAETZEL and ANDERSON, 2005). However, defining its approach and its eventual application can vary depending on who handles it or who operates its properties.

For civil engineering, for example, soil can mean a body that can be excavated, while for agricultural engineering, this same layer of unconsolidated material can indicate the surface on which life develops; and, when it comes to it, in the eyes of a biologist who uses ecology and pedology, soil can be the body that influences the biogeochemical cycling of mineral nutrients and determines the different ecosystems and habitats of living beings.

Within this context of land use we have Precision Agriculture, which can be defined as an agricultural management system based on the spatial and temporal variation of the production unit and aims to increase economic return, sustainability and minimise the effect on the environment of the indiscriminate use of correctives, fertilisers and pesticides (BRASIL, 2012). This is why Precision Agriculture can be understood as a form of crop management that takes into account the spatial variability of soil attributes (YNAMASU and BERNARDI, 2014).

With this tool, there is a partial reduction in the use of inputs per area, fungicides, insecticides and herbicides are applied at exactly the right time and, above all, in exactly the right doses, avoiding any kind of waste. These factors reduce farming costs and often result in increased production per area (ROSSATTO, 2010).

Precision Agriculture consists of a cycle of analyses of the soil's productive potential by assessing its characteristics through the collection of samples and satellite images, precise control of the application of inputs and soil correction, as well as precise control of planting and the application of agrochemicals (Figure 1). The advantages of its application are many, such as: saving on agricultural inputs, pesticides, fertilisers, agricultural correctives; increased productivity due to the optimisation of soil resources; long-term sustainability of the land, exploiting it in an optimised and non-depredatory way (PETILIO et al., 2007).

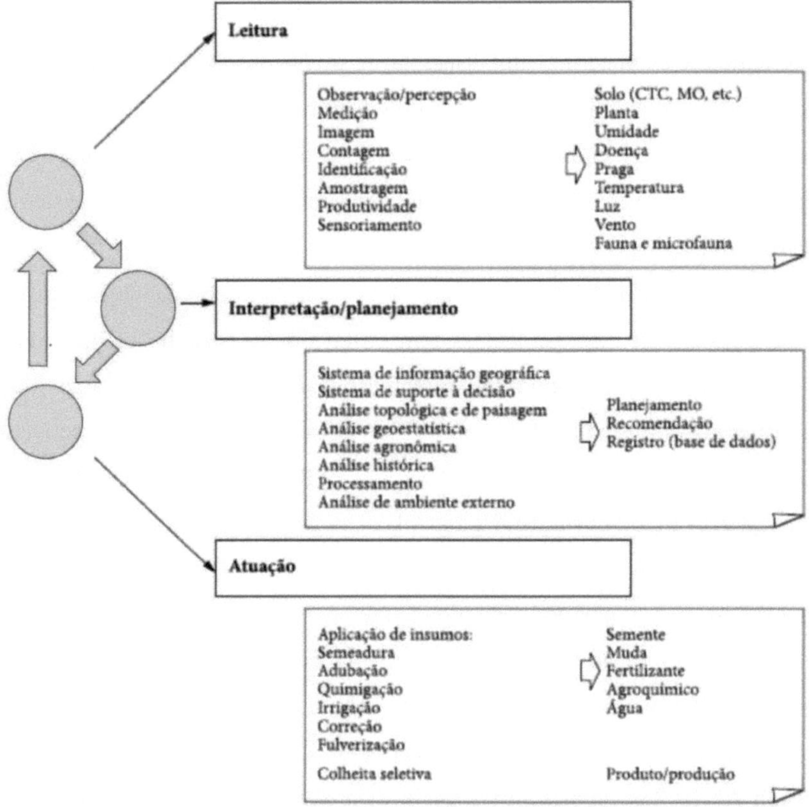

Figure 1. Cycle of precision agriculture (YNAMASU and BERNARDI, 2014).

Among the services to be carried out on the farm are mapping areas, collecting soil samples, analysing sampling data and producing maps and analysing soil samples. One of the main actions of Precision Agriculture is variable rate soil fertilisation. This means correcting with exactly the amount of fertiliser required for each field (ROSSATTO, 2010).

Variable rate soil fertilisation is carried out once every 2 or 3 years, or on an ad hoc basis whenever necessary. Maintenance fertilisation, or replacing the nutrients extracted from the previous crop, is carried out according to the characteristics of the next crop to be planted. The principle is that each small part of the crop will receive the amount of product/input it needs and no longer the amount based on the average demand of a large area, as is the case in conventional agriculture (ROSSATTO, 2010).

Due to the lack of specific information about the soil being worked, many producers fertilise the soil without a detailed analysis of the real fertilisation needs of each part of the crop. The practice of using nutrient formulations in fixed quantities, which is quite common among producers, can lead to an imbalance in the supply of nutrients. This is because certain planting areas may be receiving excessive amounts of fertiliser without the plants really needing it, or they may need more fertiliser. Precision agriculture helps to adjust the amount of fertiliser that each planting area needs (BERNARDI et al., 2015).

More precisely, this study aims to assess soil fertility in a conilon coffee plantation located at the Federal Institute of Espírito Santo - Alegre *Campus,* using Precision Agriculture techniques.

REFERENCES

BALASUBASHINI, M. S. Ferulic acid alleviates lipid peroxidation in diabetic rats. Phytotherapy Research, v. 18, p. 310-314, 2004.

BERNARDI, A. C. C.; BETTIOL, G. M.; GREGO, C. R.; ANDRADE, R. G.; RABELLO, L. M.; INAMASU, R. Y. Precision agriculture tools as an aid to soil fertility management. Cadernos de Ciência & Tecnologia, v. 32, p. 205-221, 2015.

BRAZIL. Ministry of Agriculture, Livestock and Supply. Ordinance No. 852 - Art. 1 Creating the Brazilian Precision Agriculture Commission - CBAP. Official Gazette of the Federative Republic of Brazil, Brasília, DF, 21 September 2012. Section 1, n. 184. Available at: <http://www.cnpt.embrapa.br/biblio/do/p_do42.htm>. Accessed on: 15/03/2018.

CINTRA, R.M.G.; MANCINI-FILHO, J. Antioxidant activity of spices: Evaluation and comparison of in vitro and in vivo methods. Nutrire. v.22, p. 49-62, 2001.

DEL CASTILLO, M. D.; AMES, J. M.; GORDON, M. H. Effect of roasting on the antioxidant activity of coffee brews. Journal of Agricultural and Food Chemistry, v. 50, p. 3698-3703, 2002.

ENCARNAÇÃO, R. O.; LIMA, D. R. Coffee & human health. Brasília: EMBRAPA Café, 2003.

ENVIRONMENTAL SYSTEMS RESEARCH INSTITUTE (ESRI). ArcMap/ArcGIS Desktop: Release 10. Redlands, CA: Environmental Systems Research Institute. 2011.

FAZUOLI, L. C. Genetics and coffee tree improvement. In: RENA, A. B.; MALAVOLTA, E.; ROVHA, N.; YAMADA, J. (Eds). Coffee tree culture: factors affecting coffee tree productivity. Piracicaba: POTAFOS, 1986.

FERRÃO, M. A. G.; FONSECA, A. F. A.; VERDIM FILHO, A. C.; VOLPI, P. S.; Origin, Dispersion, Taxonomy and Genetic Diversity of *Coffea canephora*. IN: FERRÃO, R. G.; FONSECA, A. F. A.; BRAGANÇA, S. M.; FERRÃO, M. A. G.; MUNER, L. H. D.; Café Conilon. Vitória: Incaper, 2007.

MOREIRA, R. F. A. Discrimination of brazilian arabica green coffee samples by chlorogenic acid composition. Archivos Latinoamericanos de Nutricion. v. 51, p. 95-99, 2001.

NACIF, A. P. Coffee and human health. Brasília: EMBRAPA Café, 2003.

PETILIO, A.; PEREIRA, M.; PERÃO, G.; TAMAE, R. Y. A brief study of the feasibility of applying precision agriculture techniques. Electronic Scientific Journal of Agronomy. Available at: <http://www.revista.inf.br/agro11/artigos/anoviedic11-art09.pdf>.
Accessed on: 15 March 2017.

ROSSATTO, T. Precision agriculture. Supervised Curricular Internship Report. Farroupilha Federal Institute - Júlio de Castilhos Campus, 2010.

SANDERS, M. E. Overview of functional foods: emphasis on probiotic bacteria. International Dairy Journal, v. 8, p. 341-347, 1998.

SCHAETZEL, R.; ANDERSON S. Genesis and geomorphology. Cambridge: Ed. Cambridge, 2005.

SILVA, C. R. M.; NAVES, M. M. V. Vitamin supplementation in cancer prevention. Revista de Nutrição, v. 14, p. 135-142, 2001.

YNAMASU, R. Y.; BERNARDI, A. C. C. Precision Agriculture: results of a new look. Available at: <https://www.embrapa.br/agriculturadeprecisao>. Accessed on: 15 Apr. 2018.

CHAPTER 1

Marc Ferrez/Moreira Salles Institute Collection

LITERATURE REVIEW

There is no denying the existence of a plethora of artistic, literary and scientific works about coffee and the way in which drinking it has transformed human relationships and social interaction. This topic compiles some of them in order to provide a brief review of coffee's origins, nutritional properties and market potential. As well as aspects related to Precision Agriculture.

1.1 The origin and spread of coffee

The true origin of coffee is not known for sure and there is no evidence of its actual discovery, but there are, however, numerous legends about its possible origins. The most widely accepted theory tells the story of the shepherd Kaldi, who possibly lived in the region of Absinia - now Ethiopia - over a thousand years ago.

1.2 The legend of Pastor Kaldi

The legend goes that Kaldi, on observing the individuals in his herd more closely, noticed that his goats became happier and more bouncy whenever they chewed on certain reddish-coloured fruits attached to the bushes surrounding the field. The shepherd also noticed that it was only after eating the fruit that his flock was able to walk long distances and overcome endless climbs. Hearing the comments about the behaviour of Kaldi's animals, a local monk took a handful of the reddish fruits with him to the monastery and noticed, after preparing a drink with them through their infusion, that the recipe helped him resist sleep while praying and during his long hours of reading the breviary. Coffee was then cultivated in various Islamic monasteries in Yemen as the effects of its consumption were reported (NEVES, 1974).

1.3   The first coffee crops

To this day, coffee plants can be found scattered throughout Ethiopia's natural vegetation, but it was Arabia that was really responsible for spreading its culture around the world, with the intensification of the consumption of *qahwa,* the "wine" of Arabia; one of the first ways of processing the fruit.

Although manuscripts dating back to 575 AD have been found in Yemen about its use in natura, it wasn't until the 16th century, in Persia, that coffee beans were roasted - and consumed in the form of the drink we know today - and it wasn't until the middle of 1615 that the drink began to be savoured on the European continent. It wasn't until 1699 that the Arab monopoly on coffee cultivation came to an end with the advent of Dutch experimental plantations in Java. The profits from the success of this experiment were so significant that they aroused the interest of the Germans, the French and the Italians (NEVES, 1974).

NEVES (1974) also says that it was only a matter of time before coffee reached the colonies of the New World. After all, the act of preparing it in a coffee maker and enjoying it socially had become the prerogative of status and refinement and, therefore, it was up to the regime of the newly conquered colonies to operationalise its production and finally supply the growing European consumer market.

1.4   Coffee in Brazil

Coffee was brought from French Guiana to Brazil - more precisely to Belém - in 1727 by Sergeant Major Francisco de Mello Palheta, at the request of the governor of Maranhão and Grão Pará. Climatic conditions led to its spread across the states of Bahia, Rio de Janeiro, Minas Gerais, São Paulo and Paraná and its production mainly served the domestic market (TAUNAY, 1939).

It only took a short period of time for coffee to occupy first place in the ranking of the basic products of the Brazilian economy, developing with total independence and with the support of national resources (TAUNAY, 1939).

The coffee plantations that were first established in the Paraíba Valley in 1825 led the country's economy into a new economic cycle. For almost half a century, wealth flowed unimpeded through the coffee plantations, occupying valleys and mountains, bringing together large urban centres - such as the south of Minas Gerais, the interior of São Paulo and the north of Paraná - boosting inter-regional trade, opening the borders to large numbers of immigrants and forever uniting the drink with the culture of the Brazilian people (DPASCHOAL, 2006).

1.5   From the 1929 crisis to the present day

29 October 1929 marked the history of the world economy forever. The crash of the New York Stock Exchange destabilised countless economies around the world, and among them was the coffee industry.

During this period, millions of stored bags were burnt and millions more coffee trees had to be eradicated in an attempt to halt the continuous fall in prices caused by surplus production. The southeast of the country only began to grow again once the world economy was back on its feet (TAUNAY, 1939).

Today, Brazil is the largest coffee producer on the planet and contributes 30% to the international market, an amount equivalent to the sum of what the other six largest exporters produce. It is also worth pointing out that while arabica coffee production is concentrated in the states of São Paulo, Minas Gerais, Paraná, Bahia and part of Espírito Santo, robusta coffee is mainly grown in Espírito Santo and Rondônia (DPASCHOAL, 2006).

REFERENCES

DPASCHOAL, L. N. Aroma de Café. DPaschoal, 2006.

NEVES, C. A estória do café. Rio de Janeiro: Brazilian Coffee Institute, 1974.

TAUNAY, A. E. História do café no Brasil: no Brasil Imperial 18221872. Rio de Janeiro: National Coffee Department, 1939.

CHAPTER 2

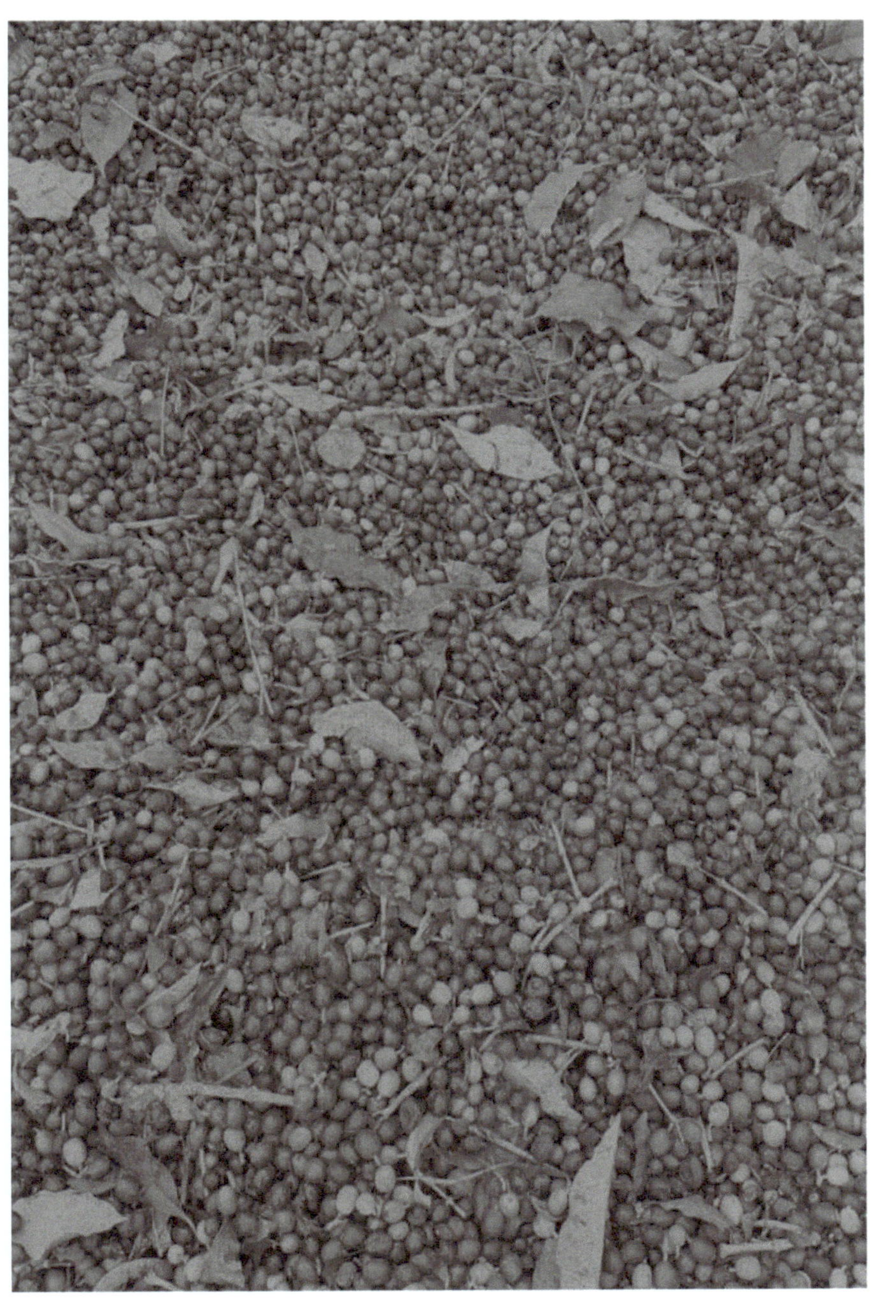

2.1 Taxonomy, classification and characteristics

According to CONAGIN and MENDES (1961), conilon coffee (C. *canephora)* is diploid with 2n = 22 chromosomes and multiplies through cross fertilisation. This incompatibility is gametophytic and linked to the S1, S2 and S3 alleles.

VALLAEYS (1952), quoted by DUBLIN (1964), tested the method of asexual reproduction using cuttings and found positive results. In other places, such as Madagascar, using individual selection of mother plants in plantations, it was possible, through the method of asexual propagation, to acquire a significant increase in the productivity of the plantations formed (SNOECK, 1968).

PAULINO et al. (1984), in Brazil, proved these results in C. *canephora,* Conilon cultivar, pointing out that it is individually favourable for this type of proliferation, as it has the advantage of being multi-stemmed, allowing the acquisition of a large number of cuttings per plant, which are simple to root.

It is an allogamous species, self-incompatible, perennial, shrubby and woody stemmed, made up of populations expressing great variability, with highly heterozygous individuals (CONAGIN and MENDES, 1961; BERTHAUD, 1980; FONSECA, 1999). Due to the natural form of reproduction, the populations varied in size, fruit shape and colour, seed size, reaction to biotic and abiotic factors and productivity.

In conditions of higher temperature and rainfall, the plants reach greater heights; the leaves are larger and less intensely green in colour than those of C. arabica; the flowers are white, in large numbers per inflorescence and per leaf axil; and the fruit varies in shape and number depending on the genetic material (CARVALHO, 1946; FAZUOLI, 1986).

2.2 Coffee's economic importance

Until the beginning of the 21st century, coffee was grown on approximately

300,000 properties, occupying an area of 2.7 million hectares in 1,850 municipalities in twelve states across the country. It is a sector of the Brazilian economy that generates around 8 million direct and indirect jobs. According to data from the 4th Coffee Harvest Survey 2016, by the National Supply Company (CONAB, 2016), the Brazilian harvest totalled 51.37 million bags (60 kg) of coffee, with the main producing states being: Minas Gerais, Espírito Santo, São Paulo, Bahia, Rondônia, Paraná, Rio de Janeiro, Goiás and Mato Grosso, which account for 98.6% of national production.

With the worldwide increase in the plant's production by less expressive countries, production ended up being greater than consumption, leading to a significant reduction in the price of coffee. CONAB (2018) reports that coffee production in Brazil is equivalent to 36% of world production and has been operating effectively within the country, with production forecast at 58.04 million bags in 2018. The arabica harvest will total approximately 44.33 million bags and conilon production will be 13.71 million bags.

Therefore, changes in the economy and competition for product prices have orientated the agricultural field towards a search for greater efficiency and better control of data at field level. The influence of humanitarian movements for greater conservation of natural resources and a reduction in the damage caused to the soil through poisoning and misuse is another factor influencing new behaviours in the management of agricultural production processes (JAKOB, 1999).

REFERENCES

BERTHAUD, J. L'Incompatibilité chez *Coffea canefora:* Méthode de test et déterminisme gènétique. Cofé Cacao Thé, v. 24, p. 167-174, 1980.

CARVALHO, A. Geographical distribution and botanical classification of the genus *Coffea* with special reference to the Arabica species. Separata dos Boletins da Superintendência dos Serviços do Café, 1946.

CONAB - NATIONAL SUPPLY COMPANY. Agricultural Observatory: Monitoring the Brazilian harvest. v. 5. n.2, 2018.

CONAB - NATIONAL SUPPLY COMPANY. Agricultural Observatory: Monitoring the Brazilian harvest. v.4. n.3, 2016.

CONAGIN, C. H. T. M.; MENDES, A. J. T. Cytological and genetic research on three Coffea species; self-incompatibility in *Coffea canephora*. Bragantia, v.20, p.787-804, 1961.

DUBLIN, P. Le bouturage du caféier Excelsa. Café Cacao Thé, v. 8, p. 3-16, 1964.

FAZUOLI, L. C. Genetics and coffee tree improvement. In: RENA, A. B.; MALAVOLTA, E.; ROVHA, N.; YAMADA, J. (Eds). Coffee tree culture: factors affecting coffee tree productivity. Piracicaba: POTAFOS, 1986.

FONSECA, A. F. A. da. 1999. Biometric analysis of conilon coffee (Coffea canephora Pierre). Viçosa, MG. Doctoral Thesis, Federal University of Viçosa, 121p.

JAKOB, A. A. E. 1999. Study of the correlation between soil property variability maps and yield maps for precision agriculture purposes. Campinas, SP. Master's dissertation, Faculty of Agricultural Engineering - FEAGRI, 145p.

PAULINO, A. J.; PAULINI, A. E.; MATIELLO, J. B. Preliminary observations on the formation of Coffea canephora cv. Conilon plantations through the rooting of cuttings. In: BRAZILIAN CONGRESS OF COFFEE RESEARCH. Londrina. Proceedings... Rio de Janeiro: Instituto Brasileiro do Café, 1984. p. 157-159.

SNOECK, J. La rénovation de la caféiculture malgache a partir de clones sélectionnés. Café, Cacao, Thé, v. 12, p. 223-235, 1968.

VALLAEYS, G. Le bouturage du caféier Robusta. Bulletin d'information de l'INEAC, v. 1, p. 205-228, 1952.

CHAPTER 3

Fertilising and liming the soil

Fertilising the soil is an important stage in growing any species, as each one requires specific amounts of certain nutrients. The quantities and types of nutrients needed to grow a species can be found in fertiliser manuals, but it is essential to carry out a chemical analysis of the soil, which will provide information about the soil's current capacity to supply nutrients to the plants. To do this, before fertilising, the producer should divide the land into homogeneous plots and take a soil sample from each plot.

After collection, the samples should be sent to a soil analysis laboratory. Soil collection and analysis should be planned around 4 to 5 months in advance of planting in case liming and/or fertilisation is required, as limestone needs at least 3 months to react. In the case of annual crops, the ideal time to collect the soil for analysis is at the beginning of the dry season and for perennial crops, such as coffee, the ideal time should be after the harvest (CASTRO, 2014).

Regulating soil pH is an important step for good productivity. Plants growing in soils with an acid pH are subject to limited growth due to the presence of aluminium and manganese at toxic levels and low microbial activity, as well as becoming more sensitive to active acidity (H), factors that limit root growth and make them sensitive to drought. In this situation, there is also a decrease in the rate of mineralisation of organic matter, which hinders the use of essential nutrients for plant growth, such as nitrogen, phosphorus, sulphur and molybdenum. In order to prevent a deficit in growth and nutrient utilisation, liming should be carried out (YARA BRASIL, 2018).

Liming the soil consists of using acidity correctors (the most commonly used is limestone) to regulate the pH, as well as being used as an additive for proper plant root growth as it is a source of calcium and magnesium. As well as regulating pH, liming increases crop productivity by reducing toxic aluminium (Al) and manganese (Mn) and increasing the availability of nitrogen (N), phosphorus (P), potassium (K) and sulphur (S). However, the effect of liming is not immediate in terms of regulating the soil's pH, as the corrective needs to react with the soil water, which takes between 60 and 90 days, depending on the lime's PRNT.

It is recommended that the soil be limed in conjunction with fertiliser recommendations according to the specific quantities needed for each part of the crop. The use of liming combined with fertilisation reduces the risk of environmental degradation and prevents crop damage (YARA BRASIL, 2018).

3.1 Coffee tree nutrition

In addition to photosynthesis, plants take nutrients mainly from the soil to carry out their metabolic processes. Balanced nutrition guarantees high productivity and grain quality (YARA BRASIL, 2018). The nutrients responsible for healthy crop growth are categorised into macronutrients: nitrogen (N), phosphorus (P), potassium (K), calcium (Ca), magnesium (Mg) and sulphur (S) and micronutrients: boron (B), zinc

(Zn), copper (Cu), iron (Fe), manganese (Mn), chlorine (Cl), molybdenum (Mo) and nickel (Ni) (MESQUITA et al., 2016).

When plants obtain nutrients from the soil and store them in their parts (roots, stems, branches, leaves, flowers and fruit), this is called absorption. The part of the nutrients removed from the plant's growth site, such as fruit and trunks, is called export and can also be returned to the site through the use of straw as organic fertiliser (MESQUITA et al., 2016).

When the plants begin to show symptoms of toxicity (excess nutrients) or "starvation" (nutrient deficit), production may have been completely jeopardised. The main noticeable symptoms are reduced growth, yellowing leaves and root problems (MESQUITA et al., 2016).

3.2 Nutrients and their functions

3.2.1 Nitrogen

It is the nutrient with the highest demand and the most accumulated in the coffee tree. When used in adequate quantities, it is responsible for satisfactory growth, but in excess it can jeopardise fruit ripening and size, as well as loss of beverage quality. In excess, nitrogen can also cause deficiencies in the micronutrients zinc, boron, copper and iron, and increase the likelihood of attacks by diseases (Phoma and Pseudomonas) (MESQUITA et al., 2016; YARA BRASIL, 2018).

3.2.2 Phosphorus

Phosphorus is the nutrient in greater demand in the formation phase than in the adult phase, as it is responsible for structuring the roots and wood. Phosphorus deficiency causes the leaves to lose their shine and colour, followed by necrosis, as well as damaging the plant's growth due to poor development of the root system. In excess, phosphorus can impair the absorption of copper, iron, manganese and zinc (MESQUITA et al., 2016).

3.2.3 Potassium

Unlike phosphorus, it is the nutrient most consumed by older plants. It plays an important role in photosynthesis, respiration and sap circulation and is accumulated in the fruit. It acts in the setting and filling phase of the fruit, influencing its size by increasing the accumulation of sugars. This nutrient is also involved in improving the coffee drink. It helps the crop resist drought and cold by controlling water loss through the opening and closing of stomata. A deficiency of this nutrient causes a change in leaf colour and subsequent drying out, as well as shrivelling of the fruit. In excess, it can cause calcium and magnesium deficiency (MESQUITA et al., 2016; YARA BRASIL, 2018).

3.2.4 Calcium

It is supplied to the crop through liming. This nutrient plays a fundamental role in planting the crop, as it is responsible for root development. It is also involved in increasing quality and tolerance to diseases that attack the fruit. Calcium must be present in the deeper layers of the soil in order to make the roots grow deeper, ensuring resistance to drought. It is also essential for bud development, fruit ripening and protein formation. Its deficiency is characterised by yellowing of the edges of young leaves and, in critical cases, the death of the terminal bud of young plants (MESQUITA et al., 2016; YARA BRASIL, 2018).

3.2.5 Magnesium

It plays a role in photosynthesis and is a component of chlorophyll. Deficiency symptoms begin in the older leaves and those close to the fruit with yellowing just between the veins leading to premature leaf fall, as well as reducing the plant's photosynthetic rate and consequently its size (MESQUITA et al., 2016).

3.2.6 Sulphur

It has metabolic and structural functions in proteins. It participates in root development and chlorophyll synthesis. When there is a deficit of this nutrient, the youngest leaves change colour to light green, progressing to chlorosis of the whole plant, defoliation and shortening of the internodes. Organic matter is the source of this nutrient, so its deficit is associated with a shortage of organic matter in the soil (MESQUITA et al., 2016).

3.2.7 Zinc

It is an important nutrient because it is directly associated with coffee production. Zinc is involved in plant growth and pollen tube germination, as well as influencing flower setting and fruit size. It is strongly retained by the soil's exchange complex, especially clay, which limits its absorption by the roots. When in short supply, the leaves become narrow, twisted, leathery, brittle and rough to the touch.

There is chlorosis that can develop into purple spots, defoliation and girdling (MESQUITA et al., 2016).

3.2.8 Boron

A limiting nutrient in coffee production, as well as zinc. It is also found in organic matter. It is closely linked to calcium, which limits its absorption. It is involved in cell elongation and division, limits root growth and makes the plant sensitive to drought. It plays an important role in pollen tube growth and pollen grain germination. The deficiency of this nutrient can be caused by leaching, excessive liming or excess nitrogen fertilisation, and is aggravated during the dry season. The death of the apical buds of the branches and the apex of the coffee tree, with several buds sprouting below. The new leaves become small, twisted and have irregular edges and a narrow limb. It can cause the flowers to fall and damage the coffee tree's production (MESQUITA et al., 2016).

3.2.9 Iron

It is the micronutrient most available in coffee plantation soils and therefore the one most accumulated in the plant. It is involved in respiration and the composition of chlorophyll. It is used in large quantities to ensure productive, strong and healthy plants. Deficiency is due to an excess of lime and organic matter or, in acidic soils, it can be due to an excess of manganese, which reduces its absorption, or due to waterlogging of the soil and a lack of light. Symptoms occur on young leaves through yellowing. Iron deficiency should be extremely avoided,
as it can seriously damage the grains (MESQUITA et al., 2016; YARA BRASIL, 2018).

3.2.10 Manganese

After iron, it is the micronutrient most accumulated in plants. Manganese participates in photosynthesis and can replace magnesium in some enzymes. When in excess in acidic soils, it impairs zinc absorption. A decrease in this nutrient is associated with excessive liming or soils with an excess of organic matter. Symptoms of manganese deficiency are seen in young leaves, with light green internerval regions with yellowish spots. Excess manganese causes chlorotic spots on older leaves and subsequent necrosis (MESQUITA et al., 2016).

3.2.11 Copper

It is generally not found in sufficient quantities in the soil and factors such as high nitrogen fertilisation, excessive liming, high organic matter content, heavy phosphate fertilisation and waterlogging are the causes of low levels of this nutrient. Deficiency is characterised by the secondary veins of new leaves becoming prominent and the development of chlorotic spots that lead to necrosis. In addition, the limb also curves, popularly known as "zebu ear" (MESQUITA et al., 2016).

3.2.12 Molybdenum, chlorine and nickel

These are the micronutrients that are found in good concentrations in the soil and do not show symptoms when they are deficient (MESQUITA et al., 2016).

REFERENCES

CASTRO, C. Momento Soja: the importance of fertilising well. Available at:<http://www.projetosojabrasil.com.br/momento-soja- importancia-da-adubacao-bem-feita/>. Accessed on: 26. Aug. 2014.

MESQUITA, C. M.; REZENDE, J. E.; CARVALHO, J. S.; FABRI- JÚNIOR, M. A.; MORAES, N. C.; DIAS, P. T.; CARVALHO, R. M.; ARAÚJO, W. G. Coffee manual: management of coffee plantations in production. Belo Horizonte: EMATER-MG, 2016.

YARA BRASIL. Soil and liming. Available at:

<http://www.yarabrasil.com.br/nutricao-plantas/culturas/soja/fatores-   chave/solo-e-calagem/>. Accessed on: 15 Apr. 2018.

CHAPTER 4

Precision agriculture

Farmers are increasingly concerned about the need to produce more in less space (CAMARGO, 1997). In this sense, producers have long sought to maximise crop production by modifying the application of inputs according to soil types and crop behaviour.

COELHO (2003) shows us that it was only with the advent of agricultural mechanisation that it became possible to economically manage crops over large areas with uniform application of fertilisers, herbicides and pest control treatments. However, the uniformity of treatments does not respect the natural and induced changes in soil properties, and can result in areas with too much and others with no nutrients, making it possible to add economic and environmental problems to this effective application of inputs to the crop.

New measures have been taken to increase the efficiency of planting and make it more profitable for producers through localised agricultural management, known as Precision Agriculture (PA). This new strategy is defined as an agglomeration of technologies and procedures that add up to better management of the production systems of a given crop (SHIRATSUCHI, 2001).

The PA system aims to understand what the soil really needs, based on information collected at different locations in a field, in order to provide important data to determine the exact need and location for the intervention. After obtaining the field data, a map of the area is generated on the computer, making it possible to observe the relationships between various factors through the attributes of the area.

According to JAKOB (1999) the maps that are generated make it possible to visualise the plots that can be used to determine the amount of fertiliser that should be applied in a given location, moderating overuse and the possibility of soil contamination in areas where it is not needed.

The application of fertilisers in different quantities is related to the importance of PA, which carries out limited treatment of an area based on the differences within the same crop. Before assessing fertiliser needs, there is a process of investigation and diagnosis.

In more developed countries that have adopted this practice, there is specific equipment for the localised application of inputs, most of which is highly improved, capable of transporting various products separately, instantly combining the desired solution and applying the products to the precise location. However, the cost of purchasing this equipment is high for producers who want to carry out the procedure over large areas. On the other hand, producers of small plantations use the equipment through services provided by co-operatives or service providers (MOLIN and MENEGATTI, 2005).

PA in Brazil is slowly being adopted. Changing the way fertilisers have been used for many years is difficult for coffee producers. Practising a PA system involves a circuit of services. The improvements to the environment are potentially very significant, although there are generally no direct financial incentives for producers to adhere to more sustainable environmental practices.

In Brazil, soils generally have low fertility and require the application of inputs to achieve yields that are considered good. With this in mind, the main objective of PA is to manage the whole of a cultivated area in such a way that agricultural gains are maximised and the impact of agriculture on the environment is reduced.

The AP concept is generally associated with the use of high-tech equipment (either hardware or software) to assess or monitor the conditions on a given plot of land and then apply the various production factors (seeds, fertilisers, phytopharmaceuticals, growth regulators, water, etc.) according to the conditions on the ground (Figure 2).

PA takes into account the spatial and temporal variation of the soil's productive potential and the specific needs of crops, in order to increase the efficiency of fertiliser use and thus improve economic yields and reduce the environmental impact of agricultural activity (COELHO and SILVA, 2009).

The objectives of Precision Farming include:

1. The increase in farmers' income, achieved by two distinct but complementary means: reducing production costs and increasing the productivity (and sometimes also the quality) of crops.

2. Reducing the environmental impact resulting from agricultural activity is related to: controlling the application of production factors, which should be tailored to the needs of the cultivars (COELHO and SILVA, 2009).

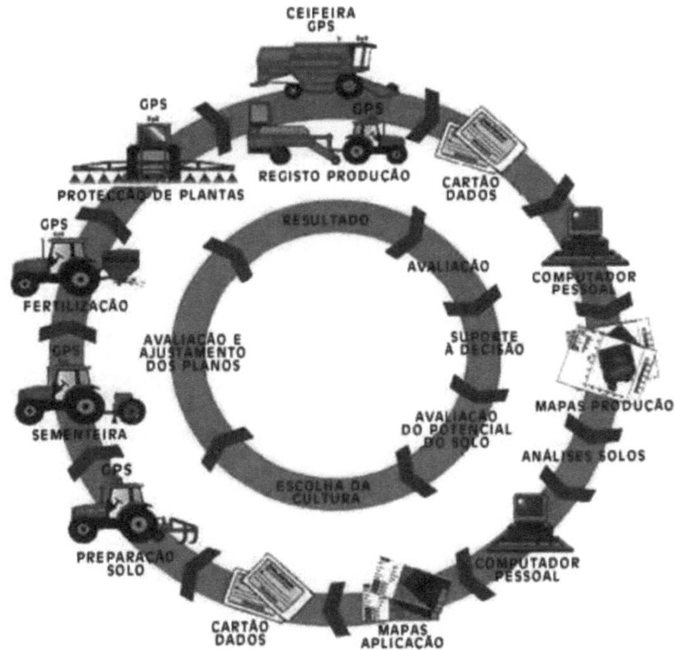

Figure 2 - General diagram of a Precision Agriculture system (COELHO and SILVA, 2009).

However, due to the high costs of this technique, it becomes more interesting if the resulting economic benefits are greater than the investment required to adopt it. Available work on PA has been developed for annual crops. However, for perennial crops such as coffee, yield mapping systems are less developed (OLIVEIRA, 2007).

According to CAPELLI (1999), the advantages of using PA are the possibility of better growth of the production field, which in a way means a better-prepared project with more informed decisions. Because it is a technique that distributes inputs in the places with the greatest need, it becomes a means of production that minimises costs and consequently increases the profitability of the crop.

There is still a long way to go for the associated information technologies, especially in terms of their ease of use and integration into the business. If the investments are to be profitable, there must be farmers and/or technicians with

33

sufficient knowledge to adjust, develop and utilise these technologies (COELHO and SILVA, 2009). In spite of everything, the future looks more favourable as high-tech equipment tends to fall in price and the educational level of farmers has increased.

Through PA it is possible to make rational use of the spatial variability of the characteristics of a plot of land. These characteristics are associated with the type of soil, such as water storage capacity, nutrient content, pH and organic matter. However, there are also characteristics that are not associated with the type of soil, such as slope, exposure to the sun and the existence of pests and/or diseases, which are also responsible for the spatial variability of crop yields (COELHO and SILVA, 2009).

In a PA system there are various tools such as Positioning Systems (GPS), Geographic Information Systems (GIS), Soil Analyses, Remote Sensing, Environmental and Productivity Monitoring Systems, Differentiated Applications (VRT - Variable rate technology) and Economic Aspects of Precision Agriculture. Among these, soil analysis deserves to be highlighted, as it is now a common practice in most agricultural production systems in developed countries.

Given the spatial variability of soils, which often reveals itself even in small plots, these analyses are essential.

When analysing soils, it is necessary to decide which are the most important variables that, under certain conditions, most affect crop growth and development. Among the main variables are:

- Fertility: content of macronutrients (N, P, K, Ca, Mg and S) and micronutrients (B, Cl, Cu, Fe, Mn, Mo, Zn and Ni);
- pH: directly related to the availability of most nutrients for utilisation by plants;
- CEC: cation exchange capacity of the soil.

With regard to agricultural soil characteristics, the following are equally important: depth, organic matter content, texture, structure, water storage capacity, internal and external drainage, permeability, compaction and cation exchange capacity. In addition, it is essential to take into account the slope and exposure of the land

(COELHO and SILVA, 2009).

If we use a GPS to determine where the samples were taken, we can know the exact location that corresponds to each soil analysis. The results of these analyses can be used to create fertility maps (in GIS), to which, among other things, different doses of fertiliser application can be associated (COELHO and SILVA, 2009).

Fertilisation is the most common application of PA systems. Fertilising the soil with manure/fertilisers or correctives is usually done to a greater extent by applying macronutrient fertilisers and lime. In conventional systems, these applications are often made using soil analyses and taking into account the potential productivity of the crop.

In PA systems, instead of using the average value resulting from the various soil samples or sub-samples taken on the plot, the specific value for each patch of soil is respected and different applications are made to each patch according to needs (COELHO and SILVA, 2009).

Coffee is one of the main agricultural products in Brazilian exports and one of the most important crops for the country's economy. In this context, in-depth knowledge of all the stages of coffee plantation management, from planting to harvesting, is of the utmost importance. Due to the diversity of factors that influence the productivity of the coffee tree, managing the crop in a homogeneous way can lead to a reduction in profitability for rural producers (FERRAZ et al., 2012).

REFERENCES

CAMARGO, E. C. G. 1997. Development, implementation and testing of geostatistical procedures (kriging) in the georeferenced information processing system (Spring). São José dos Campos, SP. Master's dissertation, National Institute for Space Research (INPE), 123p.

CAPELLI, N. L. Agricultura de precisão - Novas tecnologias para o processo produtivo LIE/DMAQAG/ FEAGRI/UNICAMP, 1999. In: TSCHIEDEL, M.; FERREIRA, M. F. Introduction to Precision Agriculture: concepts and advantages. Ciência Rural, v. 32, p. 159-163, 2002.

COELHO, A. M. Agricultura de precisão: manejo da variabilidade espacial e temporal dos solos e das culturas. In: CURI, N.; MARQUES, J. J.; GUILHERME, L. R. G.; LIMA, J. M.; LOPES, A. S.; ALVAREZ, V. H. V. Tópicos em Ciência do solo, Sociedade Brasileira de Ciência do Solo, v. 3, 2003.

COELHO, J. P. C.; SILVA, J. R. M. Agricultura de Precisão: Inovação e tecnologia na formação agrícola. Lisbon: Association of Young Farmers of Portugal, 2009.

FERRAZ, G. A. S.; SILVA, F. M.; CARVALHO, L. C. C.; ALVES, M. C.; FRANCO, B. C. Spatial and Temporal Variability of Phosphorus, Potassium and Productivity of a Coffee Crop. Engenharia Agrícola (Impresso), v. 32, p. 140-150, 2012a.

JAKOB, A. A. E. 1999. Study of the correlation between soil property variability maps and yield maps for precision agriculture purposes. Campinas, SP. Master's dissertation, Faculty of Agricultural Engineering - FEAGRI, 145p.

MOLIN, J, P.; MENEGATTI, L. Variable rate application: localised treatment. Cultivar Máquinas, Pelotas. v. 3, p.22-26, 2005.

OLIVEIRA, R. B. 2007. Mapping and correlation of soil attributes and conilon coffee plants for precision agriculture. Alegre, ES. Master's dissertation, Federal University of Espírito Santo, 150p.

SHIRATSUCHI, L. S. 2001. Mapping the spatial variability of weeds using precision agriculture tools. Piracicaba, SP. Master's dissertation, Luiz de Queiroz College of Agriculture, 116p.

CHAPTER 5

METHODOLOGY APPLIED TO DATA COLLECTION FOR PRECISION AGRICULTURE ANALYSES

The usual methodology for collecting data consists of systematising the crop in order to better characterise soil variability. To do this, the area of interest must be partitioned into different plots. The greater the partitioning (greater number of plots), the better the soil characterisation. The study was carried out in 2016 on a conilon coffee plantation located at the Federal Institute of Espírito Santo - Alegre *Campus*, Espírito Santo (20° 45' 55" S; 41° 27' 25" W), applying Precision Agriculture (Figure 3).

Figure 3 - Conilon coffee plantation (Coffea canephora) located at the Instituto Federal do Espírito Santo - *Campus* de Alegre, Espírito Santo, Brazil.

In this study, the land (Figures 4 and 5) was divided into 29 plots using AutoCAD *software* (version 2011). As soon as it was chosen as the field of study, the Garmin Etrex Vista HCx GPS was used to delimit 29 plots in the software, over a total area of 3,442 metres^2 (Figure 6). There were around 30 coffee plants in each of the plots. In addition to the plants, the free edge area of 3,908m2 was also measured (Figure 7).

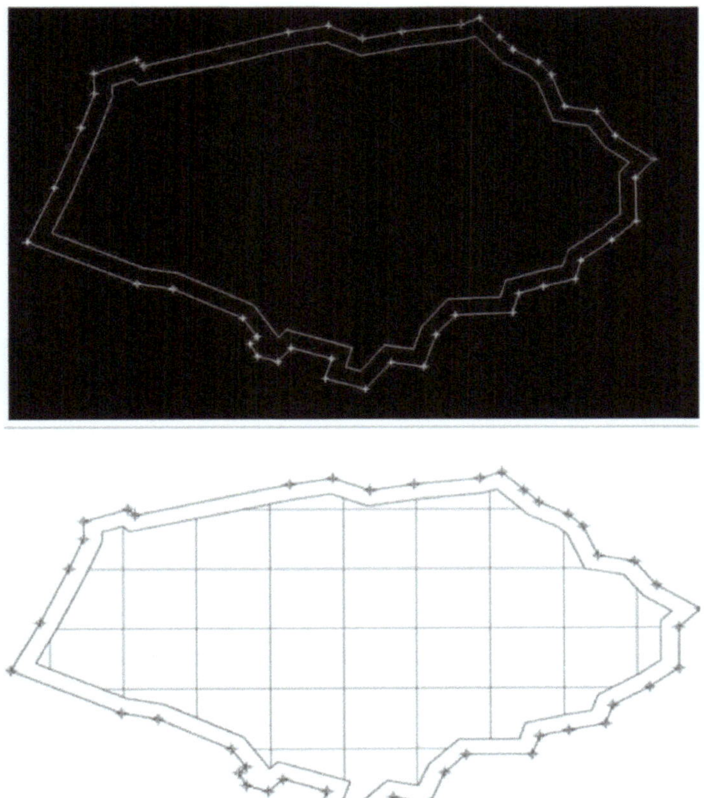

Figure 4 - Delimitation of the free edge of the terrain indicated by the (+) symbol.

The distance between plants was also measured, and the study area was chosen a priori because of its territorial dimensions. Since the area was not too large to require extensive travelling, nor too small to make it impossible, for example, to analyse more diverse soil samples.

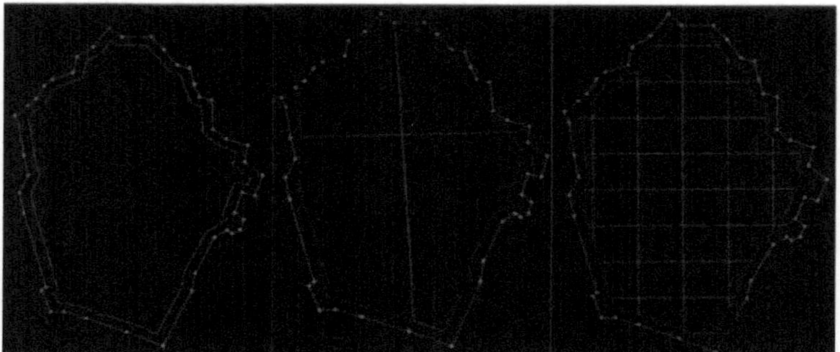

Figure 5: Division and demarcation of the plots using the AutoCAD programme.

Figure 6. Details of the work to dimension the plots in the conilon coffee plantation; Garmin Etrex Vista HCx GPS (Photo taken in the conilon coffee plantation at the Federal Institute of Espírito Santo - Alegre Campus, in 2016).

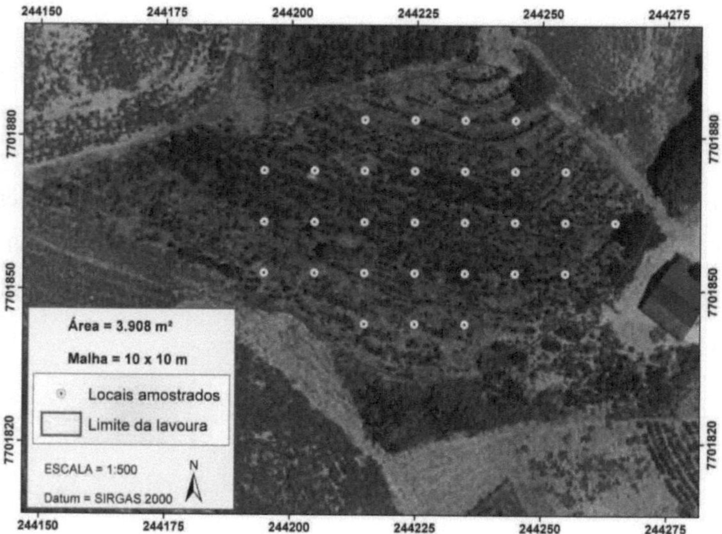

Figure 7. Geographical distribution of collection points.

5.1   Soil analyses

For the soil analyses, samples were taken from the points of intersection between the horizontal and vertical lines used to limit each of the 29 new parts of the land (Figure 5). Using a Dutch auger, 4 soil samples were collected from each of the intersection points, with the aim of mixing them to generate a single composite sample for each plot. This sampling procedure aimed to collect portions of soil that truly represented all the horizontal variability of the area under study. The samples were then sent to the CCAE - UFES Soil Laboratory for chemical analyses.

5.2   Complementary analyses

5.2.1 Soil density

The most common method for assessing soil density is the volumetric cylinder method, in which a cylinder of known volume is inserted into the layer to be analysed and removed from the soil (FERREIRA, 2010). In the example of the study carried out in 2016 on the coffee plantation located at the Federal Institute of Espírito Santo - Alegre Campus, another 4 samples were taken at the same intersection points in order to check the soil density of each of the 29 areas (Figure 8). The soil samples collected were then dried in an oven at 105°C for a period of 24 hours (Figure 9) and, using the relationship between the mass of dry soil and the volume of the cylinder, the density was obtained using the following formula $DS = Ms/Vt = g/cm^3$, where DS represents the density of the soil; Ms the mass of dry soil and; Vt the total volume of soil in the cylinder.

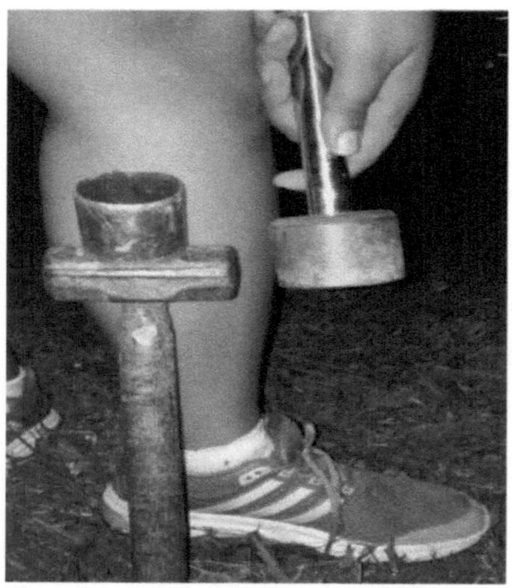

Figura 8.   Details of collection and laboratory analyses; castellated sampler used to assess soil density.

Figura 9.   Material from the 29 points taken to the oven for drying to calculate soil density

5.2.2 Statistical analyses

The data was initially evaluated using descriptive statistics, generating maps of the nutrient concentration in each plot selected in the sampling unit. The computer application ArcMap/ArcGIS (ESRI, 2011) was used to generate the maps, considering only those soil attributes with coefficient of variation values greater than 30 per cent (PIMENTEL - GOMES, 2009).

REFERENCES

ENVIRONMENTAL SYSTEMS RESEARCH INSTITUTE (ESRI). ArcMap/ArcGIS Desktop: Release 10. Redlands, CA: Environmental Systems Research Institute. 2011.

43

FERREIRA, M. M. Physical characterisation of soil. In: LIER, Q. J. van. (Ed). Soil physics. Viçosa: Brazilian Society of Soil Science, 2010.

PIMENTEL GOMES, F. Curso de estatística experimental. Piracicaba, SP: FEALQ, 2009.

PROCAFÉ. Procafé Foundation. Average reference standards for evaluating the results of soil analysis and foliar analysis in coffee growing   : <http://www.fundacaoprocafe.com.br/laboratorio/solos-e-   sheets/standards-referential>. Accessed on: 11 Apr. 2018.

CHAPTER 6

RESULTS AND DISCUSSION

In the study in question, the variables with coefficient of variation values greater than 30 per cent were: phosphorus (P), calcium (Ca), magnesium (Mg), potential acidity (H+Al), sum of bases (SB) and effective CTC (t).

The results of the descriptive analysis of the P, Ca and Mg contents and the values for potential acidity (H+Al), sum of bases (SB) and effective CEC (t) are shown in Table 1. The minimum value found for P was 3.59 mg/dm^3 and its maximum value was 148.05 mg/dm3; for Ca the minimum value was 1.73 cmolc/dm3 and the maximum 5.51 cmolc/dm3; for Mg the minimum value was 0.28 cmolc/dm3 and the maximum 1.98 cmolc/dm3; potential acidity (H+Al) ranged from a minimum of 0.74 cmolc/dm3 to a maximum of 4.62 cmolc/dm3, while the minimum and maximum values for sum of bases (SB) and effective CTC (t) ranged from 2.91 to 7.89 cmolc/dm3, respectively (Table 1). These results show that there is variability in these variables in the different plots where conilon coffee is grown, and by analysing the maps it is possible to see where this variability is occurring.

Table 1. Descriptive statistics for the variables phosphorus (mg/dm^3), calcium (cmolc/dm^3), magnesium (cmolc/dm3), potential acidity (cmolc/dm3), sum of bases (cmolc/dm3) and effective CTC (cmolc/dm3).

	Phosphorus (P)	Calcium (Ca)	Magnesium (Mg)	Potential acidity (H+Al)	Base sum (SB)	CTC effective (t)
Average	27,882	3,227	1,236	2,648	4,984	4,984
Median	17,97	2,41	1,15	2,8	3,9	3,9
Minimum	3,59	1,73	0,28	0,74	2,91	2,91
Maximum	148,05	5,51	1,98	4,62	7,89	7,89
1st Quartile	13,49	2,17	0,99	1,82	3,65	3,65
3rd Quartile	30,11	4,43	1,6	3,3	6,64	6,64
Coefficient of variation	105,86%	41,0%	31,25%	38,70%	34,19%	34,19%

According to PROCAFÉ (2011), P content is considered low when the soil analysis indicates values below 10 mg/dm3, medium for values between 10 and 20 mg/dm3 and high for values above 20 mg/dm3. Analysing the P map, it can be seen that the southern region and some of the northern region had low and medium levels, while the other regions on the map had high levels of P (Figure 10).

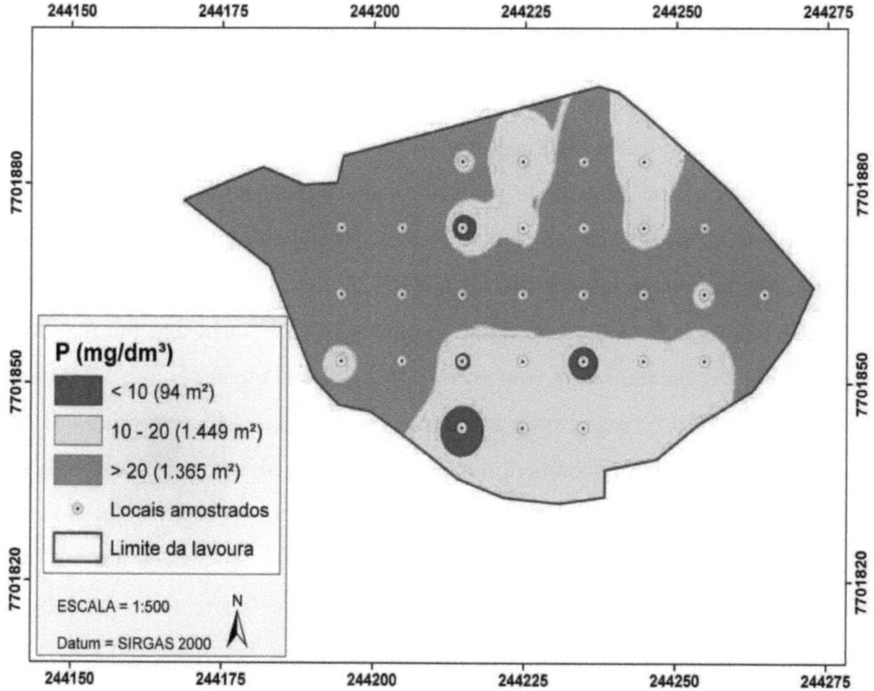

Figura 10.   Phosphorus availability in the 29 plots sampled in the conilon coffee plantation located at the Federal Institute of Education of Espírito Santo - Alegre Campus

The map shows more areas with average Ca levels, while only 11 areas located to the south of the map have good levels (greater than 4 cmolc/dm^3) (Figure 11).

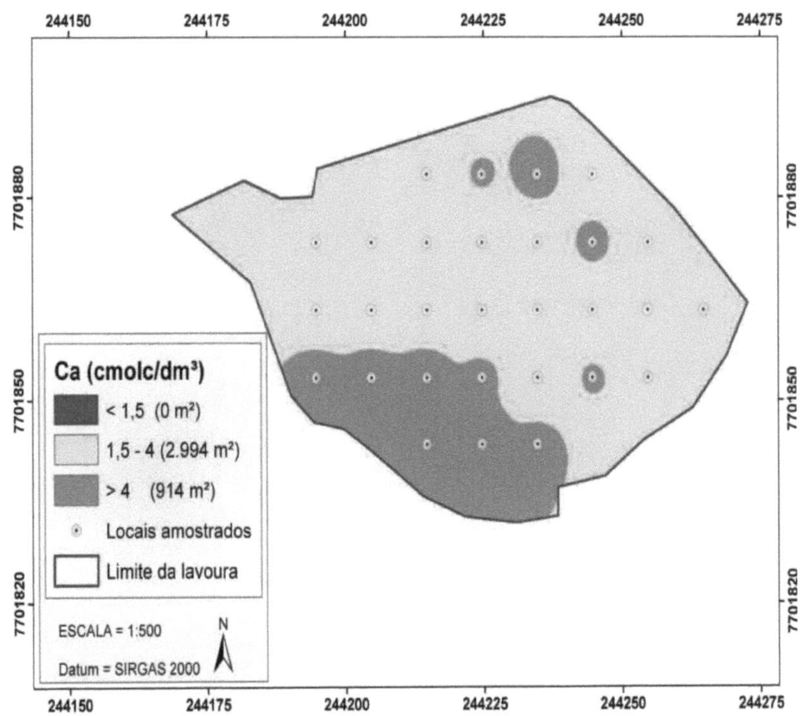

Figure 11. Calcium availability in the 29 plots sampled in the conilon coffee plantation located at the Federal Institute of Education of Espírito Santo - Alegre Campus.

As for the Mg content, the map showed a different pattern to the previous ones, with the southern region of the field not standing out. The soil content was high in most of the plots, six plots showed average levels and only one point on the map showed a low value (Figure 12).

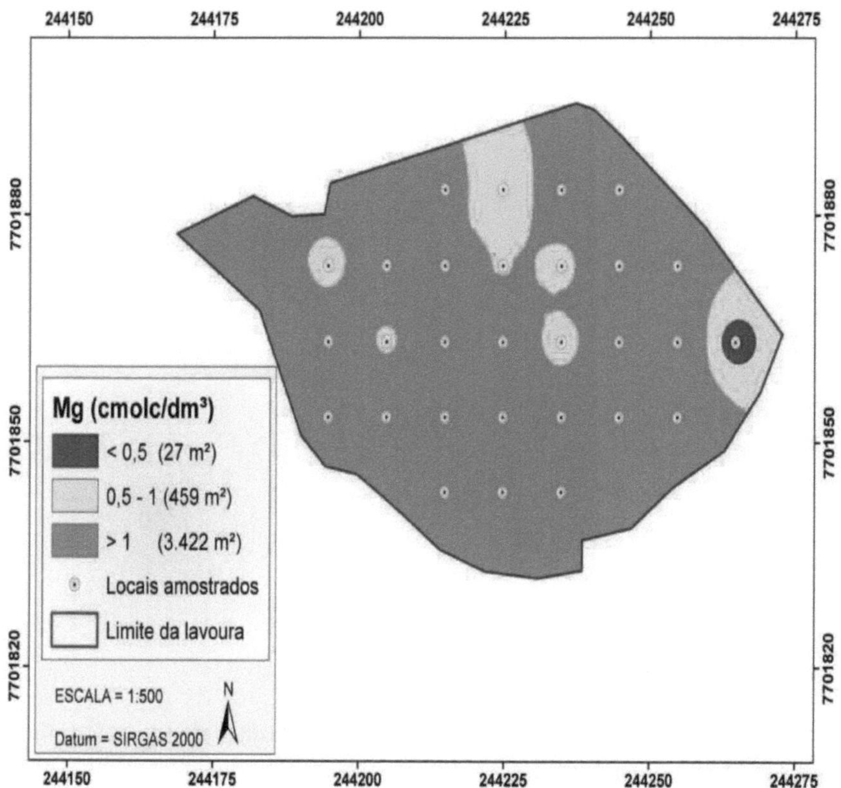

Figure 12. Magnesium availability in the 29 plots sampled in the conilon coffee plantation located at the Federal Institute of Education of Espírito Santo - Alegre Campus.

Turning to the southern region of the plantation, the maps show that most of the plots have values considered adequate for potential acidity (Figure 13). Low potential acidity values indicate a good environment for the development of the coffee tree's root system.

It was also observed that most of the plots had values considered adequate for the sum of bases (Figure 14) and effective CTC (Figure 15) for the southern region of the field. As with calcium, this is probably due to the transport of nutrients to this part of the field, as these areas are at a lower altitude than the others.

Surface transport acts to increase nutrient concentrations at lower altitudes, in crops with different degrees of altitude. In this case, the source of the nutrients is, as a rule, fertilisation carried out in the upper part of the field.

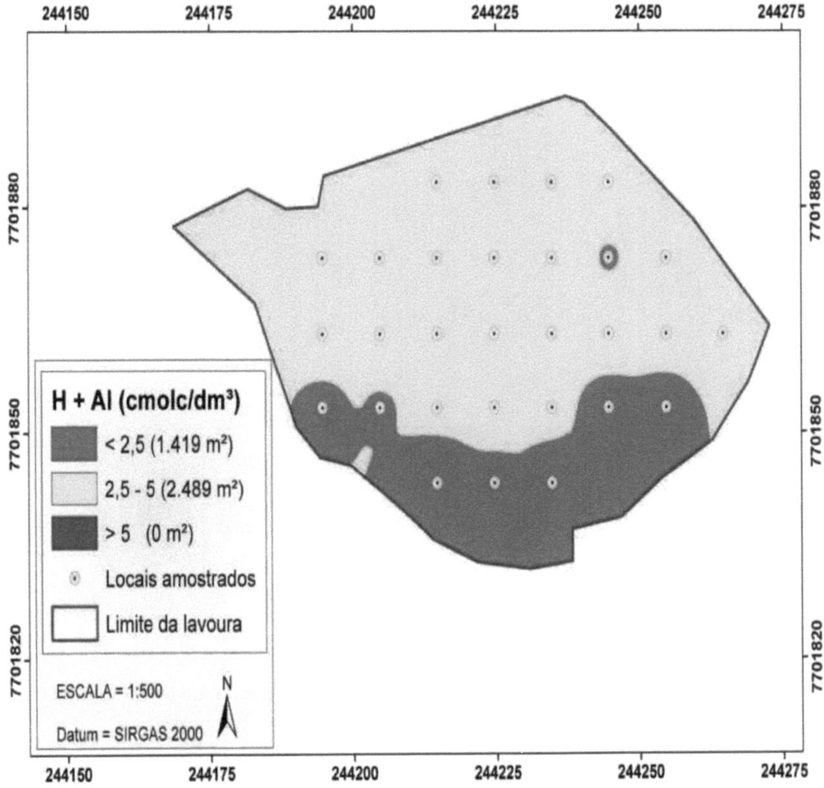

Figura 13.   Potential acidity values in the 29 plots sampled in the conilon coffee plantation located at the Federal Institute of Education of Espírito Santo - Alegre Campus

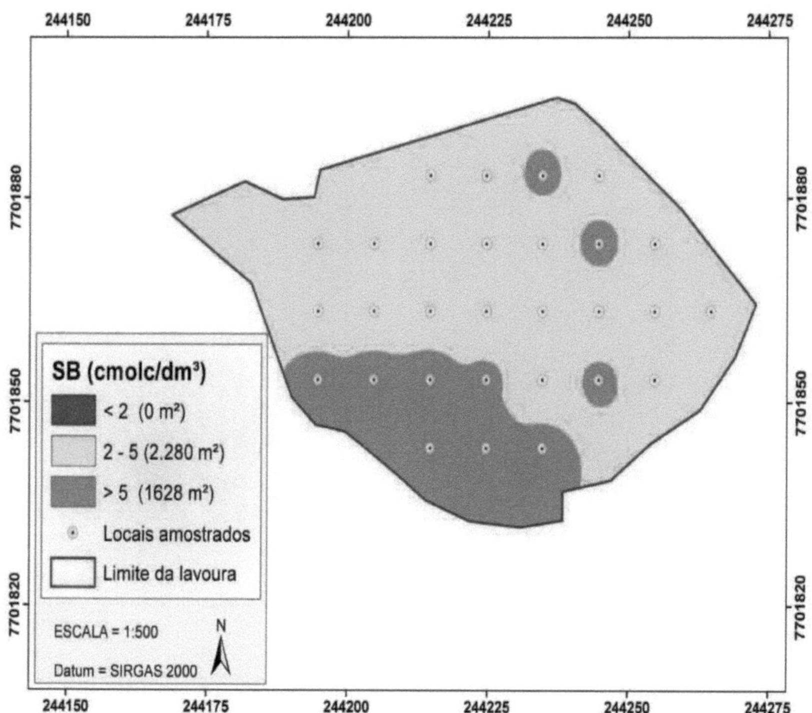

Figura 14.   Base sum values in the 29 plots sampled in the conilon coffee plantation located at the Federal Institute of Education of Espírito Santo - Alegre Campus

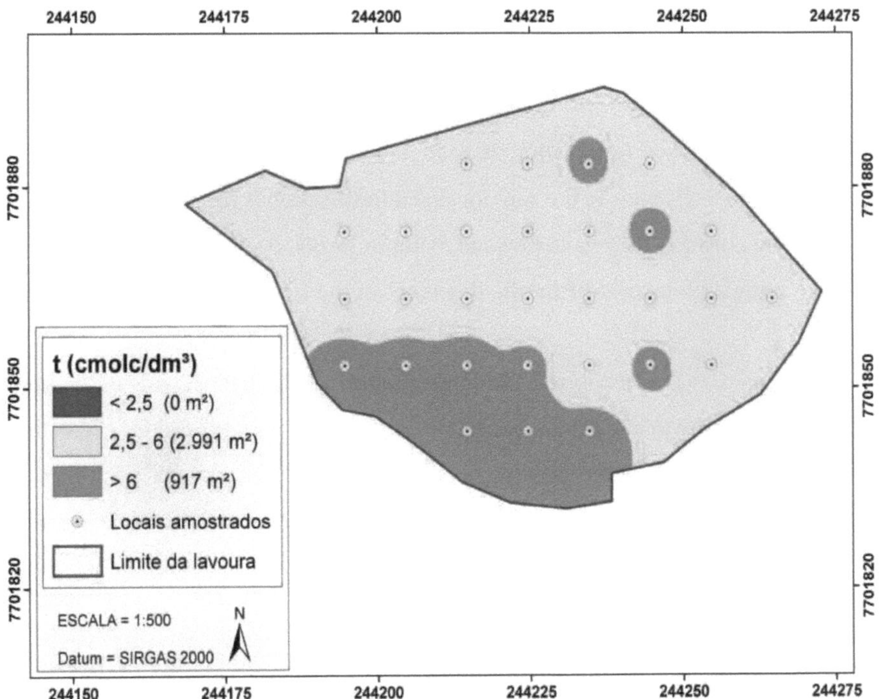

Figure 15. Effective CTC values in the 29 plots sampled in the conilon coffee plantation located at the Federal Institute of Education of Espírito Santo - Alegre Campus.

The wide range observed for the soil chemical analysis variables reveals the problems that can occur when using average values for fertility management, which is commonly used in conventional agriculture, because in some parts of the area the fertiliser application will be lower than the required dosage. In others it will be applied optimally, and in others there may be excessive application. PA has emerged as a tool for fertilising at different rates, according to the nutrient levels in each field, in order to optimise the use of fertilisers and maximise productivity, reducing production variations within a crop (FERRAZ et al., 2012b).

However, knowledge of this range alone is not enough to know where the highest and lowest levels of a given variable are to be found. Maps must therefore be produced to enable satisfactory management of the necessary interventions (FERRAZ et al., 2012b).

The values show that there is variability in the variables in the different plots where Conilon coffee is grown, and by analysing the maps it was possible to see where this variability is occurring within the twenty-nine points. In view of this, there is a need to carry out localised fertilisation on each plot.

With regard to P levels in the soil, localised fertilisation is necessary in the south and some of the north, as they have low and medium levels. As far as Ca is concerned, it should be corrected in most of the field, except in the south and some areas in the north.

According to the soil analyses, the concentration of Mg in most parts of the field is adequate for the growth and production of the crop, indicating the need for correction in one of the plots located in the eastern part of the map.

Considering the sum of bases and effective CTC, most of the plots showed values considered good to average, with the area located furthest south showing the best values.

From this study it was possible to characterise the spatial variability of the availability of phosphorus, calcium, magnesium and the values of potential acidity, sum of bases and effective CEC of the coffee crop. The availability of all the nutrients showed a structure of spatial dependence, which could only be observed through the construction of maps. Soil analysis can therefore be used in PA in coffee plantations in order to improve soil fertility and reduce spending on fertilisers, which account for a large part of the production costs of a bag of coffee.

According to the maps generated, differentiated liming and fertiliser management can be carried out. The results also show the horizontal variability of soil fertility and which attributes are most sensitive in these evaluations.

Looking at the maps for each nutrient, it is easy to see where fertiliser application is needed, highlighting the problem of using the average value for soil fertiliser recommendations. Therefore, the use of these maps has contributed to differentiated crop management, enabling more precise and efficient fertiliser applications (FERRAZ et al., 2012a).

The variability represented in the form of maps is a valuable tool for the coffee grower, as they provide useful information on where interventions should be made in their crops more precisely, such as fertilising and liming, and can therefore be considered as tools to support decision-making (FERRAZ et al., 2012b).

6.1 Soil density

Soil density, bulk density or overall density, is a physical property widely used to assess the structure of the soil.

soil. It directly reflects how the particles are arranged in the soil and their porosity. Therefore, any intervention in these particles will directly affect density (FERREIRA, 2010).

Determining the apparent density of the soil is recommended in practically all types of surveys, as it is highly relevant to various aspects of soil management (IBGE, 2007). As it is directly related to soil structure, any intervention on the particles affects density and, consequently, the development of roots will be affected. Soil density can be altered by the use and management of the soil and by changes in the arrangement of the particles, making it necessary to monitor its values over time (FERREIRA, 2010).

Changes in the soil's physical properties, such as density, indirectly affect plant development by modifying the resistance the soil offers to root growth (PEDROTTI and MELLO JÚNIOR, 2009).

With regard to soil density (DS), in the example of the study carried out in 2016 at the Federal Institute of Espírito Santo - Alegre Campus, high density values were observed. In general, it can be said that the higher the soil density, the greater its compaction and altered structure, the lower its total porosity and, consequently, the greater the restrictions on root system growth and plant development. One strategy for lowering DS is the incorporation of organic matter and the adoption of conservation soil management systems.

Table 2. Average soil density values in the 29 plots sampled in the conilon coffee plantation located at the Federal Institute of Education of Espírito Santo - Alegre Campus.

Plots	Dry soil mass (g)	Total soil volume $(cm)^3$	Soil density $(g/cm)^3$

1	130,9	88,04	1,49
2	122,73	88,04	1,39
3	113,11	88,04	1,28
4	127,15	88,04	1,44
5	127,32	88,04	1,45
6	126	88,04	1,43
7	130,62	88,04	1,48
8	127,19	88,04	1,44
9	121,84	88,04	1,38
10	125,6	88,04	1,43
11	121,58	88,04	1,38
12	120,49	88,04	1,37
13	105,22	88,04	1,20
14	106,89	88,04	1,21
15	126,34	88,04	1,44
16	126,12	88,04	1,43
17	125,64	88,04	1,43
18	116,39	88,04	1,32
19	119,17	88,04	1,35
20	119,88	88,04	1,36
21	127,07	88,04	1,44
22	132,95	88,04	1,51
23	132,59	88,04	1,51
24	128,98	88,04	1,47
25	130,73	88,04	1,4
26	116,67	88,04	1,33
27	132,87	88,04	1,51
28	132,93	88,04	1,51
29	135,83	88,04	1,54

REFERENCES

PROCAFÉ. Procafé Foundation. Average reference standards for evaluating the results of soil analysis and foliar analysis in coffee growing   :

<http://www.fundacaoprocafe.com.br/laboratorio/solos-e-referential>. Accessed on: 11 Apr. 2018.

FERRAZ, G. A. S.; SILVA, F. M.; CARVALHO, L. C. C.; ALVES, M. C.; FRANCO, B. C. Spatial and Temporal Variability of Phosphorus, Potassium and Productivity of a Coffee Crop. Engenharia Agrícola (Impresso), v. 32, p. 140-150, 2012a.

FERRAZ, G. A. S.; SILVA, F. M.; COSTA, P. A. N. da; SILVA, A. C.; CARVALHO, F. M. Precision agriculture in the study of soil chemical attributes and coffee crop productivity. Coffee Science, v. 7, p. 59-67, 2012b.

FERREIRA, M. M. Physical characterisation of soil. In: LIER, Q. J. van. (Ed). Soil physics. Viçosa: Brazilian Society of Soil Science, 2010.

IBGE. Coordination of Natural Resources and Environmental Studies. Technical manual of pedology. 2. ed. Rio de Janeiro: IBGE, 323 p., 2007.

PEDROTTI, A.; MELLO JÚNIOR, A. V. Advances in soil science: soil physics in agricultural production and environmental quality. São Cristóvão: Editora UFS, Aracaju: Fapitec, 2009.

Printed by Books on Demand GmbH, Norderstedt / Germany